Number and Operations

What Is Recycling?

Contents

What is Recycling?

Did you know that **recycle** means to use **materials** over again?

Before you put trash in the garbage, think about how it could be reduced, reused, or recycled.

The symbol for recycling shows items turning into something else.

Have you ever used the back of a piece of paper for something else? Maybe you used a water bottle for flowers. These are examples of recycling.

How Much Trash?

People produce trash, or waste, because of everyday activities.

Did You Know?

Each person in the United States produces about 5 pounds of waste per day!

Each person produces an average of 1,800 pounds of trash every year. How much trash does your whole class make in a year?

$$\begin{array}{r} 1{,}800 \\ \underline{\times\ 28} \\ 50{,}400 \end{array}$$

× 28 ← class of 28 students

50,400 pounds of trash per year!

What is Trash?

Look at the chart below. It shows the most common trash items. Notice that food scraps are third on the list.

2005 Total United States Waste Generation—245 Million Tons (before recycling)

- Paper 34.2%
- Yard trimmings 13.1%
- Food scraps 11.7%
- Plastics 11.9%
- Metals 7.6%
- Rubber, leather, and textiles 7.3%
- Glass 5.2%
- Wood 5.7%
- Other 3.4%

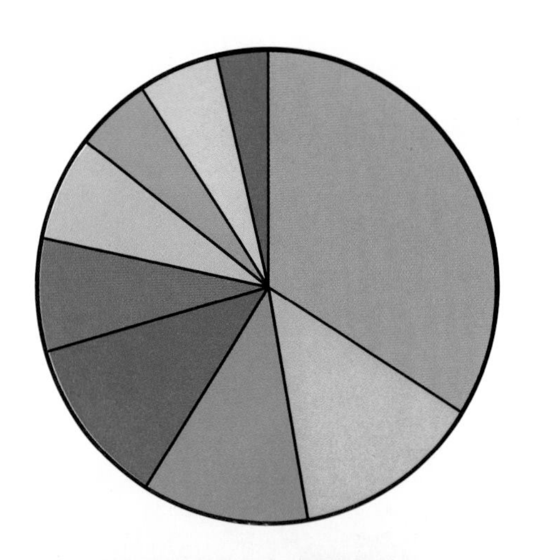

Some areas have a special recycling program for food scraps. In these areas, people set aside their food scraps in a container. Then the scraps are put into an outdoor waste bin. Trucks collect these bins. The food scraps are recycled.

What Happens to Trash—And Why Should We Care?

What happens to used aluminum cans and paper products? Most trash goes to **landfills**. Some is burned. Some is dumped into the oceans.

The number of landfills in the United States has gone down in the last 20 years. But that number does not tell the whole story. The remaining landfills have grown much larger.

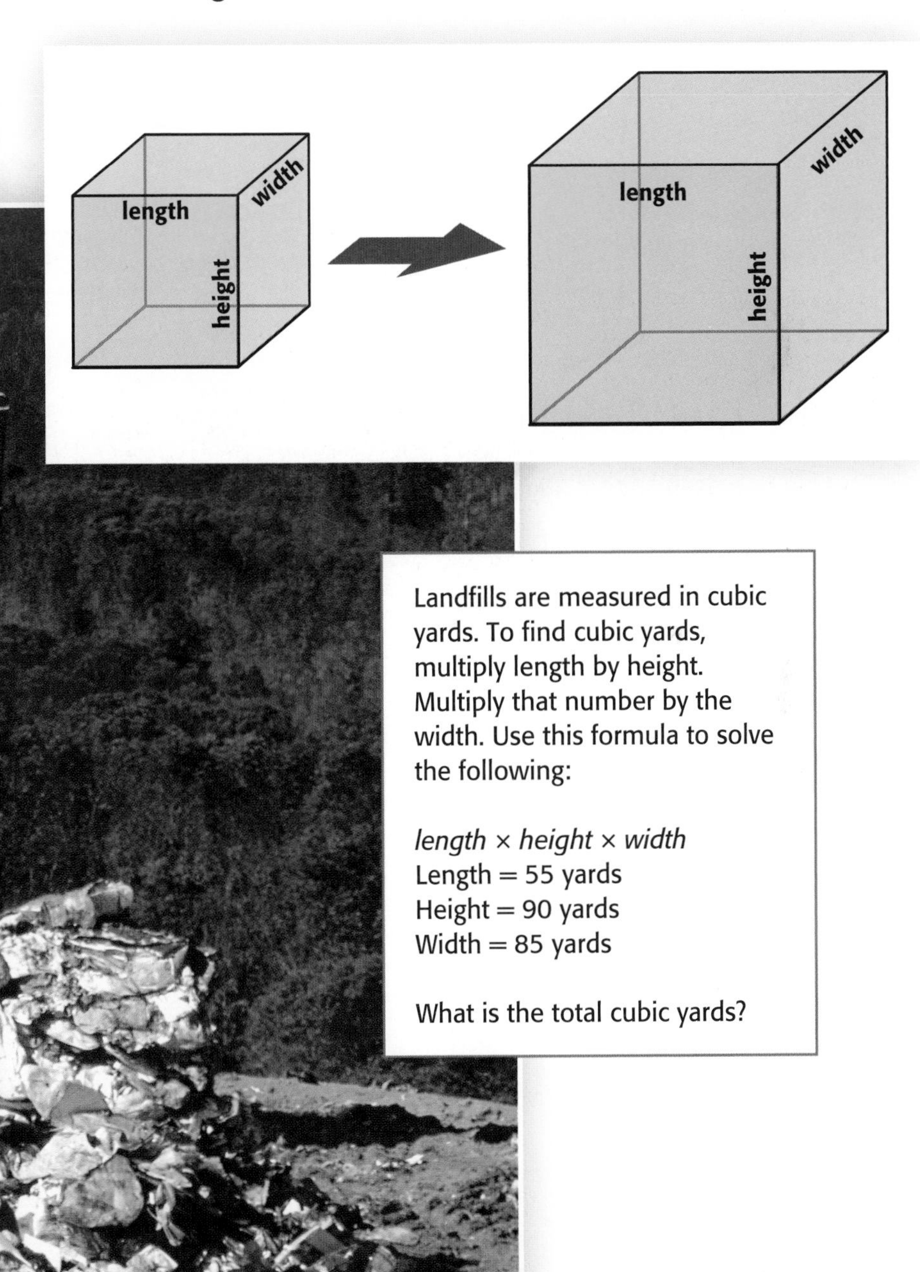

Landfills are measured in cubic yards. To find cubic yards, multiply length by height. Multiply that number by the width. Use this formula to solve the following:

length × height × width
Length = 55 yards
Height = 90 yards
Width = 85 yards

What is the total cubic yards?

Landfills can cause problems. When trash rots, it lets go of **toxins**. The toxins pollute the ground, air, and oceans. Humans, wildlife, and plants are all affected by toxins.

The United States now recycles about 5 times more trash than in 1980. About 16 million tons of trash were recycled in 1980.

Look at the graph. How much trash was recycled in 2005?

Reduce, Reuse, Recycle

The word **reduce** means to throw away less. This does not mean that you must use less of every product. Choosing products with less packaging helps reduce the amount of materials.

Not all food is trash. Banana peels do not have to be thrown away. You can use them for something else. Food scraps can be used in gardens as fertilizer. Reusing the peels helps reduce trash.

Food scraps reduce amounts of trash we throw away. Food scraps are healthy for gardens. Most kitchen waste can be made into compost. Compost is a mixture of decaying material. It helps the soil.

Each person makes 50 pounds of kitchen waste a year. How much kitchen waste does your class make in a year?

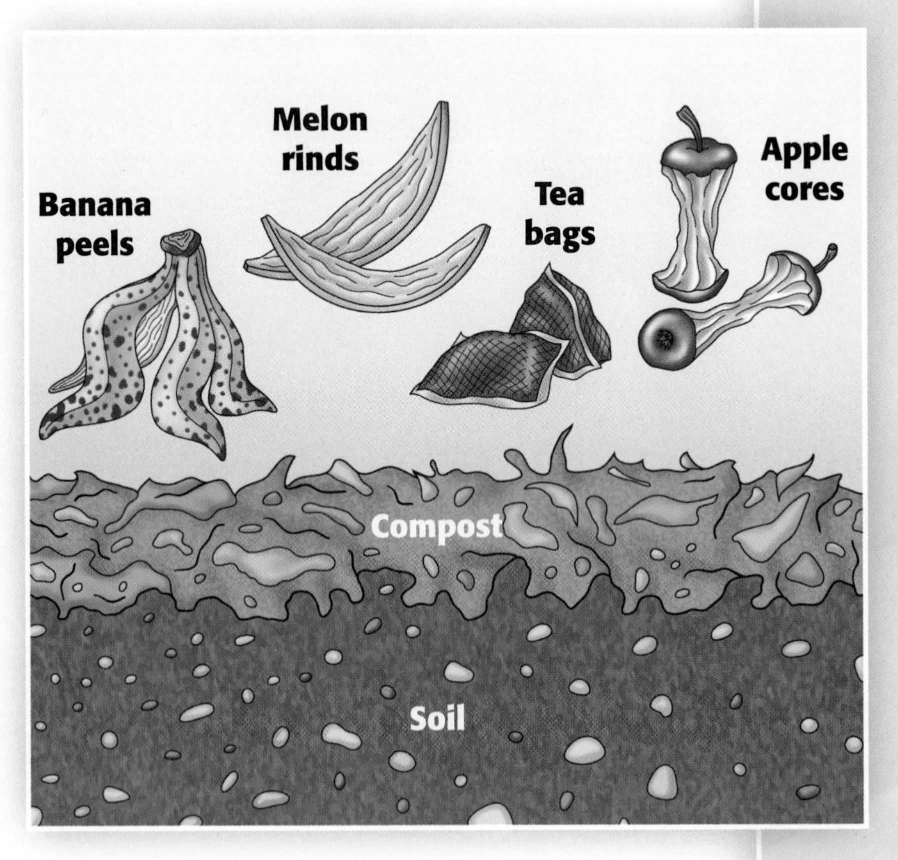

The word **reuse** means to use again. An empty milk container makes a good watering can. The container could also be used as a plant holder.

One person uses about 680 pounds of paper in one year. Each tree produces about 100 pounds of paper.

About how many trees does each person use in a year?

$\frac{680}{100} = 6.80$, or about 7 trees

It takes one ton (2,000 pounds) of recycled paper to save 17 trees. Each person uses about 680 pounds of paper each year. How many pounds of paper will your whole class save in one year?

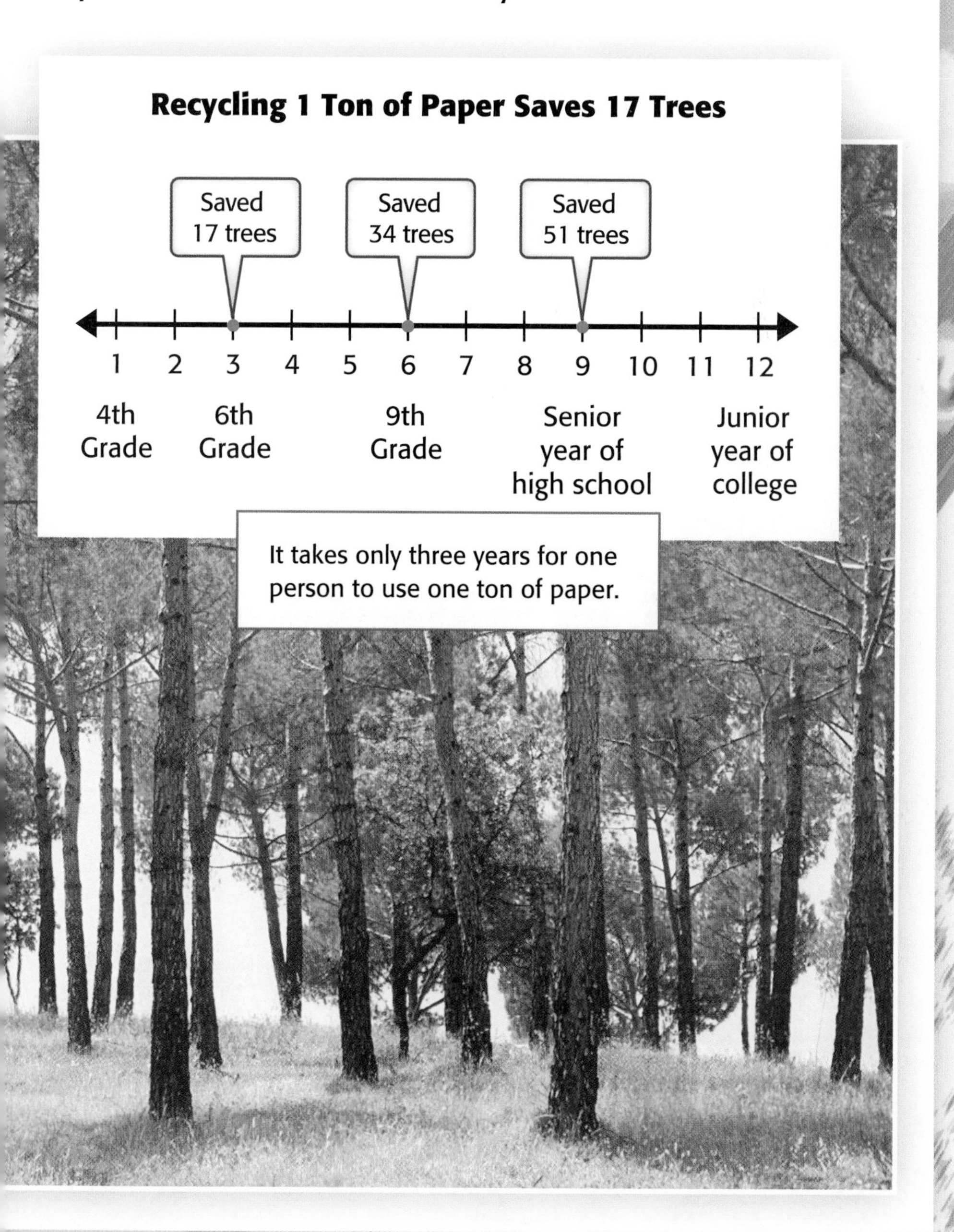

You might reuse paper in many ways. Many pet owners use old newspapers to line pet cages. They also reuse cardboard toilet paper rolls for cat toys and hamster tunnels.

You can also recycle at a recycling center. These buildings contain the items we place in recycle bins. Workers recycle the materials. Workers grind paper to make more paper. They melt soda cans to make more cans.

Recycled products can save **natural resources**. They help reduce the amount of materials in landfills. Using recycled products causes less pollution.

Did You Know?

Some playground equipment is made from different types of recycled plastic waste. The easily pressed recycled material is made into playground equipment. The rest is used as plastic chips for the playground.

What Can I Do?

You can do your part to recycle. Recycling 34 aluminum cans saves 1 pound of recyclable materials!

The typical family each year drinks about:

180 gallons of soda
30 gallons of juice
105 gallons of milk
25 gallons of bottled water

That is a lot of containers that may go to a landfill. Estimate how many gallons of bottled water a typical family would drink in 10 years.

Count how many recyclables you use for the next week. How many of these items do you use in a week, a month, and a year?

The next time you start to throw one of these items into a trash can, stop. Think: ***Reduce, Reuse, Recycle!***

	1 Week	× 4 = 1 Month	× 12 = 1 Year
Pieces of paper			
Aluminum cans			
Plastic bottles			
Glass bottles			
Plastic packages			
Cardboard			

What do the numbers 4 and 12 in the table represent? How do you know?

Glossary

landfill

A section of land used to dispose of trash. *(page 8)*

materials

Anything used for building or making something else. *(page 2)*

natural resources

Materials that are found in nature and that can be used by people in many ways. *(page 20)*

recycle

A process that allows things to be reused. *(page 2)*

reduce

To make less in amount. *(page 12)*

reuse

To apply something in a new or different way. *(page 15)*

toxins

Poisons. *(page 10)*